ISBN 978-1-5284-2230-7
PIBN 10901979

1 MONTH OF
FREE
READING

at

www.ForgottenBooks.com

By purchasing this book you are eligible for one month membership to ForgottenBooks.com, giving you unlimited access to our entire collection of over 1,000,000 titles via our web site and mobile apps.

To claim your free month visit: www.forgottenbooks.com/free901979

English
Français
Deutsche
Italiano
Español
Português

www.forgottenbooks.com

Mythology Photography **Fiction**
Fishing Christianity **Art** Cooking
Essays Buddhism Freemasonry
Medicine **Biology** Music **Ancient
Egypt** Evolution Carpentry Physics
Dance Geology **Mathematics** Fitness
Shakespeare **Folklore** Yoga Marketing
Confidence Immortality Biographies
Poetry **Psychology** Witchcraft
Electronics Chemistry History **Law**
Accounting **Philosophy** Anthropology
Alchemy Drama Quantum Mechanics
Atheism Sexual Health **Ancient History**
Entrepreneurship Languages Sport
Paleontology Needlework Islam
Metaphysics Investment Archaeology
Parenting Statistics Criminology
Motivational

Historic, Archive Document

Do not assume content reflects current scientific knowledge, policies, or practices.

Pacific Northwest
Research Station

Research Paper
PNW-RP-489
June 1996

arge- ᴗᴗ ᴗ ege ᴗ ᴗᴗ ᴗ ern̄s ın North America

Donald McKenzie, David L. Peterson, and Ernesto Alvarado

Authors

DONALD MCKENZIE is a graduate research assistant, DAVID L. PETERSON is professor of forest ecology, and ERNESTO ALVARADO is research associate, College of Forest Resources, University of Washington, Box 352100, Seattle, WA 98195.

Work was done under Cooperative Agreement PNW-93-0401 between the University of Washington and the Pacific Northwest Research Station.

Abstract

McKenzie, Donald; Peterson, David L.; Alvarado, Ernesto. 1996. Predicting the effect of fire on large-scale vegetation patterns in North America. Res. Pap. PNW-RP-489. Portland, OR: U.S. Department of Agriculture, Forest Service, Pacific Northwest Research Station. 38 p.

Changes in fire regimes are expected across North America in response to anticipated global climatic changes. Potential changes in large-scale vegetation patterns are predicted as a result of altered fire frequencies. A new vegetation classification was developed by condensing Küchler potential natural vegetation types into aggregated types that are relatively homogeneous with respect to fire regime. Transition rules were developed to predict potential changes from one vegetation type to another because of increased fire frequency. In general, vegetation currently associated with warmer or drier climates could replace existing vegetation in most biomes. Exceptions are subalpine forests and woodlands at the Arctic treeline, which are predicted to become treeless. The transition rules provide an ecological perspective on possible new configurations of vegetation types, a set of constraints for steady-state models, and a potential method of calibration for dynamic models of large-scale vegetation change.

Keywords: Biome scale, Küchler types, fire effects, transition rules, vegetation change.

Summary

Potential vegetation types are constrained by many factors, including climate. Driving variables for regional- or biome-scale simulators of vegetation change are climatological. A disturbance such as fire, however, may preclude the realization of this potential by periodically returning vegetation to an earlier successional stage. Thus, models of vegetation change in response to global warming need to incorporate the effects of disturbance at broad spatial scales. The purpose of this paper is to aid development of the Mapped Atmospheric Plant Soil System (MAPSS) model by suggesting transitions between natural vegetation types resulting from the ecological effects of fire.

A biome-scale classification of natural vegetation was developed by condensing the Küchler (1964) potential natural vegetation types into "aggregated Küchler types" that are relatively homogeneous with respect to fire regime. Based on a synthesis of · fire-effects literature considered applicable to the spatial scale of biomes, transition rules were developed to predict potential changes from one biome type to another resulting from increased fire frequency.

Vegetation currently associated with warmer or drier climates could replace existing vegetation in most biomes. Exceptions are subalpine forests and woodlands at the Arctic treeline, which are predicted to become treeless. In some cases, transitions correspond closely to potential transitions between the physiognomic vegetation types used in the MAPSS model. In other cases, where phenology or leaf type was predicted to change without a reduction in leaf area index, no correspondence exists between transitions developed in this paper and potential transitions between MAPSS biome types.

The transition rules provide an ecological perspective on possible new vegetation configurations and a set of constraints for the steady-state version of MAPSS. They also can be used to calibrate dynamic models of biome-scale vegetation change.

The rapid climatic changes anticipated for the 21st century may produce large-scale shifts in vegetation. Because of the unprecedented rate of temperature change, 10 to 50 times that of the global historical average (Schneider 1989), knowledge of current and historical patterns, although instructive, will be inadequate to predict future patterns (Brubaker 1988, Payette and others 1989). Simulation models will therefore be essential for realistic projections. These models must both address broad spatial scales and incorporate the driving processes of vegetation change at scales at which those processes can be quantified (Ehleringer and Field 1993, King and others 1990).

Potential vegetation types are constrained by many factors, including climate (Woodward 1987, Woodward and McKee 1991). Regional- or biome-scale simulators of vegetation change are driven by climatological variables, and may incorporate data from general circulation models (Neilson 1992, Running and Hunt 1993). A disturbance such as fire, however, may preclude the realization of this potential by periodically returning vegetation to an earlier successional stage. Therefore, simulation models of vegetation change, at any spatial scale, should project a different vegetation physiognomy in the absence of disturbance than in its presence (for example, Keane and others 1989, Neilson 1992).[1]

Interactions between climate and vegetation, and disturbance and vegetation are bidirectional. Vegetation composition influences atmospheric moisture and microclimate, and although fire initiation is a stochastic function of climate, vegetation also can influence fuel loading and fire severity. Existing models for predicting the effects of fire on vegetation fall into three categories: (1) mechanistic fire behavior models, (2) successional models incorporating fire stochastically, and (3) landscape-scale models of disturbance. Models in these three categories operate on different temporal and spatial scales.

Modeling Fire Behavior, Fire Effects, and Succession

Mechanistic models of fire behavior calculate, for a single fire, output variables such as flame length, fireline intensity, and scorch height from measures of fuel loading, fuel moisture levels, and wind speed (Andrews 1986; Rothermel 1972, 1991). In combination with fire effects models (for example, Keane and others 1994, Peterson and Ryan 1986, Ryan and Reinhardt 1988), fire behavior models predict stand structure and composition resulting from fire, given specific initial conditions (for example, leaf area index, species composition by percentage of basal area, and crown heights), but do not project ecological effects into the future.

Fire succession models simulate structural and compositional changes in vegetation over time on a fixed-size plot, incorporating fire initiation as a stochastic element (Keane and others 1989, Kercher and Axelrod 1984). Based on the JABOWA-FORET class of "gap" models (Botkin and others 1972, Shugart and Prentice 1992), successional models project individual tree growth deterministically, incorporating a mechanistic fire-behavior model that operates by reducing stand basal area and fuel loadings. Model outputs suggest that changes in fire frequency will strongly affect successional pathways (Keane and others 1989).

---

[1] Mladenoff, D.J.; Host, G.E.; Boeder, J.; Crow, T.R. 1993. Landis: a spatial model of forest landscape disturbance, succession and management. Draft report. On file with: McKenzie, Donald. College of Forest Resources, University of Washington, Box 352100, Seattle, WA 98195.

Landscape-Scale Disturbance Models

Although fire-behavior models predict fire spread, they do not account for the influence of spatial heterogeneity, or landscape pattern, on the propagation of disturbance. The effect of landscape pattern on disturbance spread has been simulated on abstract and real landscapes (Baker 1989, Marsden 1983, Turner and Romme 1994; also see footnote 1). Disturbance is initiated at individual pixels stochastically, with its likelihood a function of time since last disturbance or site "vulnerability." Spatial extent is usually a function of initial intensity and the vulnerability of adjacent pixels. In these broad-scale models, final burning patterns are emphasized rather than mechanistic behavior or explicit spread rates (Green 1989, Turner and Romme 1994). Connectivity of landscapes, and resulting constraints on disturbance, can be characterized by using concepts from percolation theory (Stauffer 1985).

In these broad-scale models, disturbance intensity and effects have not been simulated mechanistically at the resolution of current fire-behavior models. Rather, disturbance is characterized by generic variables such as frequency, intensity, extent, and duration. See Kessell (1976), however, for an early attempt to link fire-behavior models and landscape gradients.

Biome-Scale Disturbance—The MAPSS Model and Fire Effects

The Mapped Atmosphere Plant Soil System (MAPSS) simulates potential biosphere impacts from climate change by characterizing the dominant phenology and leaf type at a site, and then calculating the maximum leaf area that could be supported, within the constraints of the abiotic environment (Neilson 1992). Vegetation types are considered to be homogeneous over broad spatial scales.

Comparison of MAPSS output with the distribution of potential vegetation in the United States, as envisioned by Küchler (1964), produced the greatest discrepancy for the Prairie Peninsula (Neilson 1992). Application of the current MAPSS fire rule in simulations reduced this difference, thereby suggesting that fire effects can be modeled successfully at the biome scale. The current fire rule in MAPSS reduces leaf area index (LAI) based on the results of three calculations:

1. Fuel moisture and loading

2. Application of the Rothermel (1972) fire behavior model

3. Estimated crown and cambial kill (Peterson and Ryan 1986) from this output

Ecological considerations provide constraints on the vegetation transitions produced by these reductions of LAI, and additional insight into successional dynamics and changes in species composition. For example, fire can cause changes in the dominant plant phenology or leaf type in a biome without obligatory changes in LAI. Understanding how ecological effects preset the initial conditions to which fire behavior models are applied will be a key step in integrating biotic and abiotic processes, to effectively model the mechanisms of change at the biome scale.

This paper aids the development of the MAPSS model by suggesting transitions between natural vegetation types resulting from the ecological effects of fire. Rationale for the transitions comes from literature on fire effects is relevant to broad spatial scales. These transitions are applicable to a steady-state, or equilibrium version of MAPSS, in which successional dynamics are not modeled explicitly. Specifically, the objectives are to:

1. Develop a biome-scale vegetation classification for North America, delineated by geography, dominant vegetation, and distinct characteristics of fire regime. This classification is based on an existing classification and condensed to provide a rough correspondence to MAPSS categories. Condensation is based on similarities in fire ecology among vegetation types in the existing classification.

2. Suggest possible transitions among these modified types resulting from predicted changes in fire regimes under global warming scenarios and qualitatively assess the strength of the supporting evidence for the proposed transitions. It is assumed that fire frequency will increase with warmer global temperatures (Price and Rind 1994). Logic for the transitions is based on a synthesis of available literature.

3. Apply one set of transition rules to the natural vegetation of the conterminous United States.

oach

Many vegetation classifications exist for the conterminous United States (for example, Eyre 1980, Küchler 1964). We chose the Küchler (1964) classification of potential natural vegetation, because it is currently used as a basis for comparison to MAPSS output (Neilson 1992).

Aggregated Küchler Types

There are 116 potential natural vegetation types in Küchler's classification, too many, in our opinion, for establishing transition rules. In addition, many of these types are geographic variants of the same or similar dominant types. For example, there are five variants of the ponderosa pine[2] type, ranging from the American Southwest to the Black Hills, and several forest and savanna types in which oak (*Quercus*) species are dominant. Küchler also describes many geographic variants of short, tall, and mixed-grass prairie, and of mesquite savanna. If current fire regimes were similar within each group of variants, all members of the group were condensed into an aggregated Küchler type.

A secondary rationale for aggregating vegetation types was similarities in fire regime among types not necessarily homogeneous in composition of dominant species. Geographically disjunct Küchler types often have similar fire regimes, particularly in the Western states, depending on congruences of elevation or broad-scale topography. Disjunct types were combined if they were expected to respond similarly to anticipated changes in fire regime. Examples include the new aggregated types: Southeast wetlands, Western oakwoods, and Eastern deciduous forest (appendix table 7).

Transition Rules

Characterization of a fire regime must consider fire severity, fire frequency, the variability of both of these, and the importance of anomalous events (for example, crown fires in a system that usually experiences surface fires). Including all possible fire regime changes in the development of transition rules would be too complex, thereby making the results inherently subjective. Fire frequency is therefore the driving factor in predicted transitions. Although the concept of fire frequency does not directly apply to a steady-state model, such as the current version of MAPSS, it facilitates links among fire, climate, and vegetation because (1) there is better documentation for vegetation change resulting from changes in fire frequency than for other aspects of fire

---

[2] Scientific names of species mentioned in the text are given in the appendix (table 8).

regime, and (2) global warming scenarios have been more strongly linked to changes in fire frequency than to other fire characteristics, although the strongest evidence for this link comes from individual systems, and may not be applicable at the continental scale (Price and Rind 1994, Qu and Omi 1994). See also Bergeron and Archambault (1993) for opposing evidence. These transitions are intended primarily as constraints, as opposed to driving elements, in the current MAPSS fire subroutine.

Potential transitions were considered if the target vegetation type was geographically proximate to the initial type. This boundary condition was established to set limits on speculation and to keep the task manageable. When subsets of the initial type were spatially disjunct, target types were required to be proximate to only one subset of the initial type. No potential transition was rejected, however, solely because of geographical distance. For each potential transition remaining after this first step, the following questions about changes in vegetation were considered sequentially to ascertain the level of confidence with which regional or biome-scale changes might be predicted:

1. Are changes in vegetation due to increased fire frequency documented? If so, are the processes involved applicable to the spatial scale of biomes?

2a. Does fire regularly reset succession? If so, and dominant cover types change through succession, is it appropriate to assume that the physiognomy resulting from increased fire frequency will be a cross section from an earlier stage of succession?

2b. Are fire-adapted species a crucial factor in analyzing this interrupted succession? Will specific fire adaptations become more important with increased fire frequency? Will fire-induced changes in nutrient cycles or other indirect effects determine the new dominants?

3. Are the fire effects reversible, or does an entirely new successional pathway develop that is different from what would exist in the absence of fire? This is particularly important when considering ecotones. In some cases, transitions can be hysteretic; that is, decreased fire frequency would not necessarily move the transition in reverse. This is crucial for transitions, such as a change from Douglas-fir to oak woodlands in the West, that are assumed to be likely because the reverse has occurred with fire exclusion.

4. Can transitions be inferred from data on similar systems or from more general considerations? The supporting literature is often anecdotal or suggestive of possible changes, rather than rigorously documenting changes. Can possible transitions be deduced from individual species' response to more frequent fires? For example, does fire resistance or sprouting ability change significantly with age?

Application of Transition Rules

To demonstrate the procedure and the resolution level of the analysis, fire transition rules were applied to the vegetation of the conterminous United States. A geographic information system (ARC-INFO) was used to combine a vector coverage of the original Küchler (1964) types into the aggregated Küchler types described in this paper (table 1). One-step transitions were applied to each of the resulting polygons to provide a geographical display of potential changes from increased fire frequency. When more than one target type was possible, the most likely transition for each polygon was determined from local, or smaller scale, considerations.

Table 1—Aggregated Küchler types and corresponding original Küchler types

Vegetation type	Küchler identification
Spruce-hemlock	1
Hemlock/Douglas-fir	2
Silver fir/Douglas-fir	3
Western fir-spruce	4 7 15 20 21
Mixed conifer	5 29
Redwood	6
Lodgepole pine	8
Pine-cypress	9
Ponderosa pine	10 11 16 17 18 19
Douglas-fir	12 14
Cedar-hemlock-pine	13
Great Basin pine	22
Pinyon-juniper	23 24
Alder-ash	25
Western oakwoods	26 28 30
Mesquite savanna	27 60 61 62 85 87
Oak-juniper (cypress)	31 32 83 86
Chaparral ·	33 34 35 36 37
Great Basin shrubland	38 39 40 55 56 57
Desert shrubland	41 42 70 71
Desert grassland	43 44 45 58 59
Tallgrass prairie	67 72 74 75 76 77
Mixed-grass prairie	47 48 50 51 53 66 68 69 72 79
Shortgrass prairie	54 63 64 65
Grassland-wetland	49 73 78 80 92
Oak savanna	81 82 84 88 90
Blackbelt (oak-gum-cypress)	89
SE wetland (forested)	91 105 113 114
Eastern spruce-fir	93 96 97
Conifer bog (tamarack-spruce)	94
White-red-jack pine	95
Northern flood plain	98
Eastern deciduous forest	99 101 102 103
Northern conifer-hardwoods	106 107 108 109
Oak-hickory	100 104
Oak-pine	110 111
Southern mixed forest	112
Southeast pine	115 116
Alpine-Arctic tundra	52
Desert	46
Boreal woodland	—
Closed boreal forest	
Aspen-birch	
Aspen parkland	

— = Canadian type only (no corresponding Küchler type).

Vegetation Transitions

The process described above produced 40 aggregated Küchler types (table 1). In some cases, there was a one-to-one correspondence between aggregated and original types, although in others, mainly nonforested types, as many as 10 Küchler types were condensed into one type. In addition, four entirely new types were created, covering some vegetation types of Canada and Alaska at a coarse scale of resolution.

Correspondences among these modified Küchler types, MAPSS types, and Society of American Foresters types (Eyre 1980) are in the appendix. Forty-four transitions are predicted as a result of increased fire frequency (tables 2-4). Most transitions are from forested vegetation types to types with either less tree cover or dominance by earlier seral species. No transitions are predicted away from any of the grassland or desert-tundra types, although fire is definitely a factor in these systems (see discussion of transitions). Each transition is assigned a mnemonic code, to identify the type of evidence used to justify it and a numeric code (1-3), to indicate the relative strength of the evidence supporting the transition (1 = strongest).

In the text that follows, initial vegetation types are assigned one of the following categories:

1. Forest (F)
2. Tree savanna (TS)
3. Shrubland (S)
4. Grassland (G)
5. Desert/tundra (DT)

Each transition is identified by its modified Küchler type and a unique letter and number code, with letters taken from the five categories above. Tables 2-4 summarize the transitions, the logic used, and the level of confidence for each. Figure 1 is a graphic representation of the transitions.

Not all transitions in descriptive classes necessarily lead to a transition in MAPSS vegetation classes, and in some cases, there are more than one MAPSS transitions. For example, the transition spruce-hemlock → hemlock-Douglas-fir (transition F1) does not entail a change in MAPSS biome classification, whereas the transition ponderosa pine (MAPSS tree-savanna-evergreen-needle) → w. oakwoods (transition F12) results in the possibility of MAPSS classes tree-savanna-mixed-warm or tree-savanna-deciduous-broadleaf.

Initial Type: Forest

Transitions are possible from one forest type to another, or to a system with fewer trees, for example, savanna or woodland-parkland. Forest → forest transitions usually involve shifts in species composition from less to more fire-adapted species, which may or may not entail significant reductions in LAI. Forest → nonforest transitions entail significant reductions in density or LAI as well.

Table 2—Potential transitions from Western forest vegetation types due to increased fire frequency

Start type	End type	Code[a]	Conf[b]	Key references
Spruce-hemlock	Hemlock/ Douglas-fir	Doc	2	Huff (1984), Agee and Huff (1987)
Spruce-hemlock	Alder-ash	Succ	3	Franklin (1988)
Silver fir/Douglas-fir	Douglas-fir	Oth	1	Agee (1993)
Hemlock/Douglas-fir	Douglas-fir	Doc	1	Agee and Huff (1987), Huff (1984)
Douglas-fir	Ponderosa pine	Succ	2	Keane and others (1989), Barrett (1988)
Redwood	Douglas-fir	Oth	1	Finney and Martin (1992)
Cedar-hemlock-pine	Ponderosa pine or Douglas-fir	Succ	1	Fischer and Bradley (1987)
Mixed conifer	Ponderosa pine	Oth	1	van Wagtendonk (1985), Kercher and Axelrod (1984)
Douglas-fir	W. oakwoods	Doc	1	Agee amd Dunwiddie (1984), Kauffman and Martin (1987)
Douglas-fir	Shortgrass or prairie	Succ	2	Fisher and Clayton (1983)
Ponderosa pine	Shortgrass or mixed-grass prairie	Oth	2	Bock and Bock (1984)
Great Basin pine	Alpine tundra	Oth	2	Bradley and others (1992a, 1992b)
Great Basin pine	Great Basin shrub or shortgrass prairie	Succ	1	Bradley and others (1992a, 1992b), Fischer and Clayton (1993)
Pine-cypress	W. oakwoods	Oth	3	—
Fir-spruce	Lodgepole pine	Succ	1	Muir (1993), Bradley and others (1992b), Marsden (1983), and Woodard (1994)
Fir-spruce	Alpine tundra	Doc	1	Little and Peterson (1991), Billings (1969), Vale (1981), and Agee and Smith (1984)

— = no key reference.

[a] Doc = documented change with fire or change in fire-return interval; Succ = assumption of a resulting earlier successional stage; Oth = other rationale, for example, inference from knowledge of similar transitions.

[b] Conf = measure of confidence in transition logic (1 = strongest).

Table 3—Potential transitions from Eastern and Canadian forest vegetation types due to increased fire frequency

Boreal forest	White-red-jack or lodgepole pine	Doc	1	Payette (1992), Carleton and Maycock (1978)
Boreal forest	Boreal woodland	Doc	1	Morneau and Payette (1989), Sirois and Payette (1991), Foster (1985), and Sirois (1992)
Boreal forest	Aspen-birch or aspen parkland	Succ Doc	1	Payette (1992), Anderson and Bailey (1980)
E. spruce-fir	White-red-jack pine or conifer-hardwoods	Succ	1	Payette (1992), Heinselman (1981), Pastor and Mladenoff (1992), and Heinselman (1981)
Conifer-hardwoods	White-red-jack pine or aspen-birch	Succ	1	Pastor and Mladenoff (1992), Heinselman (1981)
E. deciduous (Midwest)	Oak savanna	Doc	1	Grimm (1984), Abrams (1992), and Clark (1990)
E. deciduous (East)	Oak-pine or oak-hickory	Doc	1	Abrams and Nowacki (1992), Lorimer (1985)
Conifer bog	Aspen-birch	Succ	2	Pastor and Mladenoff (1992)
S. mixed forest	Oak-pine or SE pine	Doc	1	Hartnett and Krofta (1989), Vose and others (1994), Myers (1985), and Cain and Shelton (1994)
SE wetland	Grassland-wetland	Succ	2	Christensen (1988)
Blackbelt	Oak-juniper	Oth	3	—
Blackbelt	Tallgrass prairie	Succ	1	—
Oak-pine	SE pine	Doc	2	Boerner and others (1988), Buchholz (1983)
Oak-hickory	Oak-pine	Doc	1	Hartnett and Krofta (1989)
Oak-hickory	Oak savanna	Doc	1	Anderson and Brown (1986)

— = no key reference.

[a] Doc = documented change with fire or change in fire-return interval; Succ = assumption of a resulting earlier successional stage; Oth = other rationale, for example, inference from knowledge of similar transitions.

[b] Conf = measure of confidence in transition logic (1 = strongest).

Table 4—Potential transitions from savanna and shrubland types due to increased fire frequency

Start type	End type	Code[a]	Conf[b]	Key references
Boreal woodland	Arctic tundra	Doc	1	Sirois and Payette (1991),
Aspen parkland	Mixed-grass prairie	Doc	1	Anderson and Bailey (1980)
Oak-juniper	Tallgrass or mixed-grass prairie	Doc	1	Abrams (1986, 1992)
Pinyon-juniper	Great Basin shrub or desert grassland	Doc	1	Jameson (1987), Everett (1987)
Oak savanna	Tallgrass prairie	Doc	1	Abrams (1986, 1992)
Mesquite	Desert grassland or prairie (3 types)	Succ	2	Martin (1975), Wright and Bailey (1982)
Great Basin shrub	Desert grassland	Doc	2	Humphrey (1974)
Desert shrub	Desert grassland	Doc	2	Humphrey (1974)
Chaparral	Desert grassland	Doc	2	Keeley and Keeley (1981, 1988)

[a] Doc = documented change with fire or change in fire-return interval; Succ = assumption of a resulting earlier successional stage.

[b] Conf = measure of confidence in transition logic (1 = strongest).

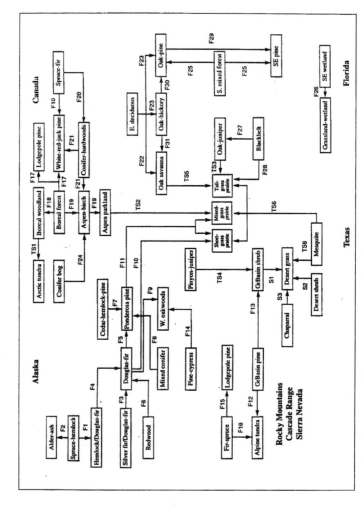

Figure 1—Transition flowchart—potential changes in natural vegetation resulting from increased fire frequency. Letter per number codes correspond to transitions discussed in text. Letter codes are as follows: F = forest, TS = tree savanna, and S = shrub.

Fire-resistant trees are characterized by thick bark, open crowns, deep roots, and large buds (Brown and Davis 1973). Many pine species have these traits, as well as adaptations such as cone serotiny or superior postfire regeneration. Some other conifers, notably Douglas-fir and western larch, become as fire resistant as pines only with age, as bark thickness increases. Thus, increased fire frequency in conifer systems may shift dominance away from late-successional species toward pines. The most fire-adapted hardwoods are oaks and aspen. The same trend away from shade-tolerant species to "early successional" species is predicted for hardwood-dominated systems with increased fire frequency. Relative dominance of pines and oaks at the endpoint of transitional pathways in forested systems may depend on local conditions such as nutrient availability after fire (Boerner and others 1988) and the ability of particular species to colonize burned sites.

F1: Spruce-hemlock → hemlock/Douglas-fir

Fire is generally not a factor in dynamics of the coastal Sitka spruce-western hemlock forest. Windthrow is the principal disturbance, and acts at small spatial scales. The presence of Sitka spruce and absence of Douglas-fir indicate a fire-return interval longer than the age of most stands, whereas presence of Douglas-fir suggests the increased importance of fire in stand dynamics (Agee and Huff 1987, Fahnestock and Agee 1983, Huff 1984). Increased presence of Douglas-fir in old-growth stands in the hemlock/Douglas-fir zone is related to increased fire frequency (Huff 1984).

F2: Spruce-hemlock → alder-ash

The Küchler alder-ash type is associated with riparian areas. In upland disturbed sites in the spruce-hemlock zone in the Pacific Northwest, alder is a strong colonizer that becomes more dominant with frequent disturbance (Franklin 1988), although this would probably require a greater increase in fire frequency than the transition to hemlock-Douglas-fir. See also F3 and F4.

F3: Silver fir/Douglas-fir → Douglas-fir

Western white pine is a successful colonizer on burned sites and may increase in Pacific silver fir zones in the Pacific Northwest with more frequent fires (Agee 1993). It is unlikely that it could dominate stands, so the transition is assumed to be to another fire-adapted species, Douglas-fir, that has this potential.

F4: Hemlock/Douglas-fir → Douglas-fir

The percentage of Douglas-fir in stands in the western slopes of the Cascade Range and Olympic Mountains is related to time since fire, with shorter fire-return intervals associated with greater percentages of Douglas-fir (Agee and Huff 1987, Huff 1984).

F5: Douglas-fir → ponderosa pine

The percentage of ponderosa pine in many forests of the Cascade Range and Rocky Mountains is related to increased fire frequency (Agee 1993, Barrett 1988, Keane and others 1989). When a seed source for ponderosa pine exists, increased fire in mixed-species stands will produce a transition to a pine-dominated system (Keane and others 1989).

F6: Redwood → Douglas-fir

Some coastal redwood stands have endured frequent, low-intensity fires, whereas others have endured moderate severity, less frequent fires. A transition in the coastal redwood zone to forests with more dominance by Douglas-fir is possible, although moisture may be more of a factor in such a transition than any change in fire frequency (Agee 1993). Unusually short fire-return intervals in redwood forests could lead to dominance by other species already present in the region (Finney and Martin 1992).

F7: Cedar-hemlock-pine → ponderosa pine or Douglas-fir

Succession in forests on the western slope of the Rocky Mountains is reversed to various degrees depending on the intensity of fires (Fischer and Bradley 1987). Ponderosa pine and Douglas-fir are dominant seral species in these forests, and increased fire frequency could reset the successional process before the "climax" stage (cedar-hemlock) is reached.

F8: Mixed conifer → ponderosa pine

The relative dominance of ponderosa pine in western mixed-conifer forests is positively correlated with fire frequency (Kercher and Axelrod 1984), and reduced fire frequency in these forests permits late-successional species such as white fir to increase (van Wagtendonk 1985).

F9: Douglas-fir → western oakwoods

The principal evidence to support this transition also comes from the reverse condition: fire exclusion in the oak savannas of Oregon and Washington has led to invasion of conifer species, particularly Douglas-fir (Agee 1993, Agee and Dunwiddie 1984, Kauffman and Martin 1987, Reed and Sugihara 1987). Thus, if a seed source for oak (particularly Oregon white oak) exists, oaks may be expected to reestablish on sites that have been invaded and dominated by Douglas-fir.

F10: Douglas-fir → shortgrass prairie

On the eastern slope of the Rocky Mountains in Montana, Douglas-fir forests form an ecotone with shortgrass prairie and invade the grasslands during atypically long fire-free intervals (Fischer and Clayton 1983).

F11: Ponderosa pine → shortgrass prairie or mixed-grass prairie

Ponderosa pine is adapted to surface fire, but a crown fire could kill an overstory (Bock and Bock 1984, Fischer and Clayton 1983), thereby converting a stand to grassland. The treeless condition would be maintained by increased fire frequency because ponderosa pine saplings would not have developed sufficient fire resistance by the time fire returned. This transition should be applied only to the ponderosa pine systems where it is documented (Black Hills and Utah woodlands). It also incorporates known variability of fire intensity into the prediction.

F12: Great Basin pine → alpine tundra

At high elevations, bristlecone pine and limber pine systems could become treeless, as the subalpine meadows in the Pacific Northwest have, when more frequent fires prevent reestablishment (Fischer and Clayton 1983). This would be more likely with increased fire intensity (Keown 1977).

F13: Great Basin pine → Great Basin shrub or shortgrass prairie

At low elevations, the same logic applies as for F12, but with different replacement species (Fischer and Clayton 1983).

F14: Pine-cypress → western oakwoods

This is a potential, but speculative, transition that is based on the fact that California oaks have greater geographic continuity and more fire adaptations than do coastal California pine and cypress *(Cupressus* spp.) The coastal pine species are serotinous, however, and a transition might be possible only if fire intensity continues to be high.[3]

F15: Western fir-spruce → lodgepole pine

Spruce-fir forests in the Rocky Mountains, Cascade Range, and high elevations in the Southwest could be replaced by lodgepole pine with an increase in fire frequency, because this pine species is shade intolerant, fire adapted (serotinous cones), and an early colonizer on burned sites (Agee 1993; Bradley and others 1992a, 1992b; Muir 1993). In areas currently dominated by red fir, however, more frequent, moderate-intensity fires could favor the more resistant red fir over lodgepole pine (Chappell 1991, Taylor 1993). In the southern Rocky Mountains, stands of aspen occur under various environmental conditions (Peet 1981). Although we expect an increase in aspen stands with increased fire frequency, the spatial scale of the modeling effort and the need to retain the Küchler system precluded consideration of an aspen endtype.

F16: Western fir-spruce → alpine (meadow-tundra)

Fires are infrequent in the subalpine zone, but damage usually is severe (Agee 1993). Reestablishment of trees after fire may be slow, particularly on sites with no seed source for lodgepole pine, with treeless areas created by fire remaining as meadows for many years, and timberline remaining below the physiological limits for tree growth (Agee and Smith 1984, Arno and Hammerly 1984, Billings 1969, Little and others 1994, Vale 1981).

F17: Boreal closed forest → pine (white, red, or jack pine — lodgepole pine in the West)

Increased fire in a mixed-boreal forest will favor fire-adapted early successional species, such as jack pine (Payette 1992 and references therein).

F18: Boreal closed forest → boreal woodland

Fire may negatively affect the productivity, and thus the potential tree density, of forests at the northern boreal ecotone, partly by providing an opportunity for lichen mat colonization and exclusion of seedlings (Morneau and Payette 1989). This is a case (see also transition F32) in which closed forest at an ecotone with nonforest is potentially unstable.

---

[3] Personal communication. 1995. J. Agee, professor of forest ecology, College of Forest Resources, University of Washington, Seattle, WA 98195.

F19: Boreal closed forest → aspen-birch or aspen parkland

The aspen parkland forms an ecotone between the boreal forest and the northern prairie in Canada. Aspen is one of the least shade-tolerant boreal species and one of the most fire tolerant due to its sprouting ability. Conifer invasion of stands pioneered by aspen results from fire exclusion (Wright and Bailey 1982 and references therein). This suggests that increased fire frequency could convert southern boreal forest in central Canada to open aspen parkland or to a forest dominated by early successional hardwoods (Payette 1992).

F20: Eastern spruce-fir → white-red-jack pine or conifer-hardwoods

This transition uses the same logic as F16. Spruce-fir forests have long fire-return intervals, thereby enabling shade-tolerant species such as balsam fir to establish. More fires would produce a transition to pines or mixed spruce-fir and hardwoods (Furyaev and others 1983, Heinselman 1981, Payette 1992, St. Pierre and others 1991).

F21: Conifer-hardwoods → white-red-jack pine or aspen-birch

The conifer-hardwood zone, sometimes referred to as the "North Woods," is a broad ecotone between the closed boreal forest to the north and the eastern deciduous forest to the south (Pastor and Mladenoff 1992). Increased fire in this region would favor fire-adapted boreal forest species such as jack pine and quaking aspen, possibly narrowing the ecotone by moving its northern edge south (Carleton and Maycock 1978, Heinselman 1973).

F22: Eastern deciduous (Midwest) → oak savanna

Mature deciduous forest burns infrequently. The principal late-successional species (for example, maple [*Acer* spp.], basswood [*Ulmus* spp.], elm, and eastern hemlock) are not fire adapted and likely would be replaced by colonizers such as oak, pine, and aspen with more frequent fires (Abrams 1992, Grimm 1984, Host and others 1987, Lorimer 1985).

F23: Eastern deciduous (East) → oak-hickory or oak-pine

Species composition in the pine-oak-hickory forest in the East and Southeast is directly related to the frequency of fire, with hickory (*Carya* spp.) being less well adapted than the other two dominant species (Abrams 1992, Bryant and others 1993, Lorimer 1985, Nowacki and others 1990, Orwig and Abrams 1994). Thus, there may be a gradual transition from the more diverse Eastern deciduous forest to a forest increasingly dominated by pine or oak, or both, with increasing fires. Transitions F22 and F23 correspond to different climatic regimes, and different available colonizers in the prairie-forest ecotone in the Midwest than in the deciduous-pine-hardwoods ecotone in the East.

F24: Conifer bog → aspen-birch

This transition is inferred from the combination of the tolerance of aspen for wet sites with its superior colonization abilities, and its preeminence, along with paper birch as an early successional species in the conifer-hardwoods zone (Pastor and Mladenoff 1992).

F25: Southern mixed forest → oak-pine or southeast pine

There is substantial documentation on the relations between fire frequency and dominant species in the Southeast (for example, Cain and Shelton 1994, Hartnett and Krofta 1989, Vose and others 1994). Systems with the most frequent fires tend to be dominated by pines, whereas those with infrequent fires support diverse forests dominated by late-successional species. Oak forests are intermediate. The logic of this transition also is applied to F29, F30, and F31.

F26: Southeast wetland (forested) → grassland-wetland

Fire resets succession in the pocosins (Christensen 1988), one of the subsets of the Southeast forested wetland type. A similar potential is assumed to exist in mangroves and cypress swamps.

F27: Blackbelt → oak-juniper

This system, unique to Alabama and Mississippi, has increased in forest cover in the last century, presumably after cessation of burning by natives and Euro-Americans (Rostlund 1957). With increased fire, it could become open savanna, with oaks, pines, or cypress juniper (*Juniperus* spp.) dominating.

F28: Blackbelt → tallgrass prairie

This transition is another possibility. Tallgrass prairie fragments are part of this type. The transition type (F27 or F28) could depend on the amount fire frequency is increased.

F29: Oak-pine → southeast pine

This transition uses the same logic as in F25. See also Buchholz (1983) and Myers (1985), but see Boerner and others (1988) for a more complex interpretation.

F30: Oak-hickory → oak-pine

This transition uses the same logic as F25. See also Quaterman and Keever (1962) and Tester (1989).

F31: Oak-hickory → oak savanna

Oak forests may be less stable than oak savannas, thus increased fire could stabilize these oak-dominated hardwood forests at lower tree densities (Anderson and Brown 1986).

Initial Type: Tree Savanna

Tree savannas are distinguished from forests by a lack of canopy closure at the mature stage. The distinctions between forest and savanna in MAPSS were retained for the aggregated Küchler types presented here. Tree savannas include parklands and woodlands at both the upper and lower latitudinal or elevational treelines. Transitions to grassland or shrubland at the southern or lower elevational ecotone may occur due to fire, even if moisture stress does not preclude tree establishment. The dynamics are complicated at the northern limit. Regeneration may be impossible even if survival of mature trees is unaffected, so fire may reduce or eliminate the time lag in the response of vegetation to climate change.

TS1: Boreal woodland → Arctic tundra

At the northern extreme of treeline, stands may not be able to reestablish after a stand-replacing fire, even though climatic conditions are unchanged (Cwynar and Speer 1991, Desponts and Payette 1993, Landhäusser and Wien 1993, Payette 1993, Sirois and Payette 1991). Thus, increased fire frequency could move the northern treeline to the south, partially offsetting any northward movement due to climatic warming (Rochefort and others 1994).

TS2: Aspen parkland → mixed-grass prairie

Prescribed fire has been used successfully to control the invasion of grasslands by aspen (Anderson and Bailey 1980). As in other tree savannas, increased fire frequency is expected to produce a transition to grassland with or without shrubs, which may or may not persist in the presence of frequent fire.

TS3: Oak-juniper → tallgrass prairie or mixed-grass prairie

The logic for this transition, as for oak savanna, is that more frequent fires will prevent savanna trees from regenerating in grassland, even where climatic conditions are favorable. Sites colonized by oaks because of fire suppression, or because the fire-return interval has been long enough to allow trees to reach a less vulnerable stage of development, will revert to a treeless condition with increased fire frequency (Abrams 1986, 1992). Grassland types will depend on geographic proximity of individual species. Many species of oak are fire tolerant, so the transition is less clear for stands dominated by oaks than for stands dominated by juniper.

TS4: Pinyon-juniper → Great Basin shrub or desert grassland

Under some conditions, usually on more mesic sites, pinyon-juniper woodlands will recover from fire (Bunting 1987, Everett 1987). On more xeric sites, recovery is not automatic, and increases in fire frequency could cause a transition to either grassland or low shrubs (Miller and Weigand 1994). Local conditions determine which pathway this would take (Jameson 1987).

TS5: Oak savanna → tallgrass prairie

Similar dynamics are expected for oak-grassland ecotones as for pinyon-juniper-grassland or oak-juniper-grassland (Abrams 1986, 1992; Tester 1989). Pinyon-juniper or oak invasions occur with fire exclusion in areas where the moisture is adequate to support woody vegetation. The logic is the same as in TS3.

TS6: Mesquite savanna → desert grassland or prairie (short, mixed, and tall grass)

Mesquite has become established on former grasslands whose fine fuels have been depleted by overgrazing (Martin 1975, Wright and others 1976). Depending on location of mesquite types (27, 60, 61, 62, 85, 87 in Küchler), the transition could be to desert or to any of the more mesic grasslands but might depend on restoration of native grasses (Wright and Bailey 1982).

Initial Type: Shrubland

Shrub-chaparral generally is fire adapted, and increased fire frequency might change community structure but not necessarily cause a transition to another type, because species composition may be more strongly determined by climate and evolutionary history, and a shift from seeding to sprouting species might stabilize the type in the presence of more frequent fire. In some systems with large shrubs that provide a substantial fuel source, increased fire may move the system toward smaller shrubs or even grasses. Particularly intense fires or short intervals between fires may result in dominance by herbaceous species with compositional changes over several decades (Christensen 1985, Zedler and others 1983). On the other hand, fire exclusion may allow competing shrubs to dominate a stand and restrict overstory regeneration (Vose and others 1994), thereby suggesting that increased fire in certain shrub systems might allow reestablishment of fire-tolerant tree species, and a possible reversal of the expected transition "arrow" in certain systems. The literature on fire effects for North America generally does not distinguish among the many shrub classes in Küchler in terms of possible succession from one type to another. Geographic restrictions may preclude the need to infer shrub → shrub transitions.

S1: Great Basin shrub → desert grassland

In dry shrubland systems in which the vegetation is still continuous enough for fire spread, grasses may gain a competitive advantage under increased fire frequency (Humphrey 1974), and woody shrubs could disappear.

S2: Desert shrub → desert grassland

This transition uses the same logic as S1 and is possible only with enough continuous vegetation.

S3: Chaparral → desert grassland

This is a possible transition, although chaparral is generally fire adapted and burns more completely after a few years of dead fuel accumulation (Callaway and Davis 1993, Keeley and Keeley 1988, Wright and Bailey 1982). Thus, the growth cycle of chaparral controls fire frequency. More frequent fires might only burn a lower percentage of biomass but maintain the system as chaparral.

Initial Type: Grassland

Grasslands generally are fire adapted. The main changes here will be in community composition, particularly in the relative abundance of C_3 and C_4 grasses in response to seasonal timing of fires (Collins and Gibson 1990, Gibson and Hulbert 1987, Howe 1994). In general, fires are more frequent in moist grasslands, whereas they do proportionally more damage in xeric grasslands (Risser 1990). It is not clear how increased fire frequency by itself might change grassland types. Transitions to desert types would more likely be driven by increased temperature or decreased moisture, or both.

Initial Type: Desert-Tundra

Desert fuels are sparse. Although fire does occur, changes in desert types are driven principally by temperature and available moisture.

Application of Transitions to the Conterminous United States

We implemented one-step transitions for all the vegetation types in the conterminous United States (figs. 2 and 3), thereby replacing the total area of the initial type. If more than one transition was possible, or there was a high probability of no transition, we selected the most likely transition based on known ecological conditions or geographic proximity. For example, western fir-spruce polygons were transformed to either alpine tundra (Olympic Peninsula and western Washington Cascade Range) or lodgepole pine (Oregon Cascade Range and central Rocky Mountains). Likewise the Eastern deciduous polygons were transformed to oak-savanna, oak-hickory, or oak-pine, depending on geographical proximity.

The most striking results of these transitions are the expansion of treeless areas at the upper treeline and the homogenization of forested types. For example, at the upper treeline in Washington west of the crest of the Cascade Range, the amount of alpine tundra is predicted to increase with the conversion of subalpine forests and parkland (transition F16). In coniferous forests west of the Cascade Range, the percentage of Douglas-fir is predicted to increase, as forests dominated by Pacific silver fir and western hemlock are converted to Douglas-fir (transitions F3 and F4) and the spruce-hemlock type is converted to hemlock/Douglas-fir (transition F1). The Douglas-fir forests in northeastern Washington and Oregon, which have significant amounts of grand fir, are predicted to become ponderosa pine forests. Table 5 shows how a particular subset, the aggregated types in Washington and Oregon, was homogenized by the one-step transitions.

Discussion

Geographic and Elevational Trends

Vegetation transitions generally were characterized by vegetation types being replaced by a new type from a more southerly, warmer or drier zone, or from a lower elevation (tables 2-4 and fig. 1). Precipitation generally decreases from north to south, and from west to east in western North America, and from east to west in eastern North America. Although figure 1 suggests only geographical trends and could be misleading for types such as ponderosa pine that have widely dispersed subsets, it indicates that patterns will be more complex and less broadly defined in the West. This concurs with results expected from the greater topographic complexity of Western biomes.

Transitions at the highest elevations (fir-spruce → alpine tundra or Great Basin pine →alpine tundra) and most northern latitudes (boreal woodland → Arctic tundra) are exceptions to the general trends. Regeneration is slow in these systems. An increase in fire frequency would reduce the window of opportunity for the establishment of late-successional or slow-growing species such as subalpine fir and black spruce. Burned sites might also be less conducive to growth and survival (Sirois and Payette 1991). If no seed source exists for a fire-adapted species, such as lodgepole pine, increased fire frequency could cause these marginal sites to revert to tundra, although the picture is complicated by predicted temperature increases that could encourage growth. Outcomes could be very sensitive to local conditions, thereby rendering predictions at broad spatial scales particularly uncertain (O'Neill and others 1989, Slatyer and Noble 1992).

Aggregated Küchler Types (pretransition)

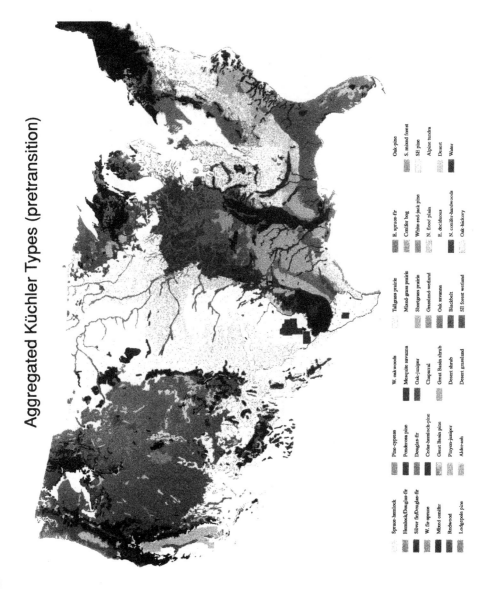

Spruce-hemlock	Pine-cypress	W. oakwoods	Tallgrass prairie	R. spruce-fir	Oak-pine
Hemlock/Douglas-fir	Ponderosa pine	Mesquite savanna	Mixed-grass prairie	Conifer bog	S. mixed forest
Silver fir/Douglas-fir	Douglas-fir	Oak-juniper	Shortgrass prairie	White-red-jack pine	SE pine
W. fir-spruce	Cedar-hemlock-pine	Chaparral	Grassland-wetland	N. flood plain	Alpine tundra
Mixed conifer	Great Basin pine	Great Basin shrub	Oak savanna	E. deciduous	Desert
Redwood	Pinyon-juniper	Desert shrub	Blackbelt	N. conifer-hardwoods	Water
Lodgepole pine	Alder-ash	Desert grassland	SE forest wetland	Oak-hickory	

Figure 2—Aggregated Küchler types in the conterminous United States before transitions were applied.

Aggregated Küchler Types (posttransition)

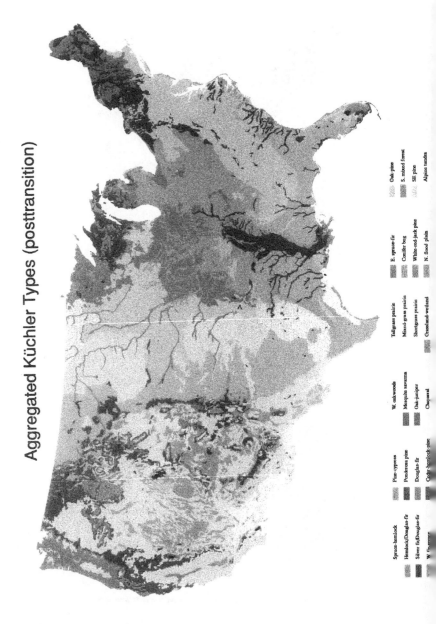

Spruce-hemlock	Pine-cypress	W. oakwoods	Tallgrass prairie	E. spruce-fir	Oak-pine
Hemlock/Douglas-fir	Ponderosa pine	Mesquite savanna	Mixed-grass prairie	Conifer bog	S. mixed forest
Silver fir/Douglas-fir	Douglas-fir	Oak-juniper	Shortgrass prairie	White-red-jack pine	SE pine
W. spruce	Cedar-hemlock-pine	Chaparral	Grassland-wetland	N. flood plain	Alpine tundra

Figure 3—Aggregated Küchler types in the conterminous United States after transitions were applied.

Table 5 —Transitions applied to the aggregated Küchler types in Washington and Oregon

Initial type	Final type
Spruce-hemlock-cedar	Hemlock/Douglas-fir
Hemlock/cedar/Douglas-fir	Douglas-fir
Silver fir/Douglas-fir	Douglas-fir
Douglas-fir	Ponderosa pine
Redwood	Douglas-fir
Cedar-hemlock-pine	Douglas-fir
Mixed conifer	Ponderosa pine
Fir-spruce (Washington west)	Alpine tundra
Fir-spruce (Oregon and Washington east)	Lodgepole pine
Great Basin shrub	Desert grassland

Correspondences to Current MAPSS Fire Rule

Fire transitions predicted by a model that produces changes in LAI, but not successional changes, can concur with transitions predicted from ecological effects only when these effects do not involve changes in phenology or leaf type. For example, the transition oak-hickory → oak savanna has an exact parallel, forest-deciduous-broadleaf-tree-savanna-deciduous-broadleaf, in MAPSS (appendix, table 7). In both cases, leaf type and phenology are unchanged, but LAI is reduced. Transitions such as Douglas-fir → western oakwoods or spruce-hemlock → alder-ash have no direct equivalent in MAPSS because they entail changes in phenology and leaf type along with substantial decreases in LAI.

Other transitions could produce an increase in LAI. A late-successional boreal forest dominated by black spruce generally will have lower productivity and less biomass than earlier stages in which trees grow more vigorously and decomposition is faster (Bonan and Shugart 1989). The transitions boreal forest → white-red-jack pine or aspen-birch will in all likelihood increase LAI. Similarly, increased fire frequency in systems currently dominated by tall, rapidly growing shrubs that have established as a result of fire exclusion might allow fire-resistant trees to reestablish, thus increasing LAI (Vose and others 1994). We are not predicting any specific transitions for this latter type, however.

A fourth type of transition entails only a compositional change, as a result of interrupted succession, and may have no counterpart in the current version of MAPSS. Although MAPSS allows fire-induced reductions in LAI to change phenology or leaf type, it has no explicit rules for succession of lifeforms. Examples are (1) spruce-hemlock → hemlock/Douglas-fir (2) western fir-spruce → lodgepole pine and (3) eastern spruce-fir → white-red-jack pine.

Additional Factors Affecting Vegetation Transitions

Although we used increased fire frequency in this paper to determine possible transitions, other aspects of fire regime clearly interact with frequency and could be either antagonistic to or synergistic with the effects of changes in frequency. Three of these additional factors are (1) season of burning, (2) variability in fire frequency, and (3) fire severity and associated anomalous events.

Season of burning—In systems with more than one potential dominant plant type, the timing of fires may control which plants are most successful. In mixed-grass prairie where both C_3 and C_4 grasses occur, the relative abundance differs depending on seasonal timing of fires (Collins 1992, Collins and Gibson 1990, Ewing and Engle 1988, Howe 1994). Although most lightning ignitions occur in July and August, prescribed burns usually are conducted in spring and may be artificially maintaining a tallgrass prairie of big bluestem, a C_4 species, in what otherwise would be a more diverse, mixed-grass prairie (Howe 1994). For similar reasons, the return of fire frequencies similar to those of the 19th century or earlier in eastern Washington and Oregon, and restoration of certain plant assemblages, will depend on seasonal timing of prescribed burns (Agee 1994).

Variability of fire frequency—In Western conifer forests, the dynamics of succession in relation to fire frequency are complex, and variation in fire-return interval may be more important than its mean (Agee 1993). For example, pathways of succession may be altered for decades in lodgepole pine-western larch forests, when variable fire-return intervals accentuate different dispersal abilities, maturation ages, and fire resistance of these species (Cattelino and others 1979). In general, in forests in which fire resistance of a potential dominant species increases dramatically with age, a single unusually long fire-free interval may have the same ecological effect over the lifetime of one cohort as an equivalent mean fire-return interval.

Fire severity and anomalous events—Infrequent severe fires, particularly in systems characterized by frequent, low-intensity fires, cause a disproportionate share of total damage (Strauss and others 1989). Crown fires have qualitatively different dynamics (Rothermel 1991), and crown-fire ecosystems have different large-scale spatial patterns (Turner and Romme 1994). Fires in these systems are often precipitated by anomalous weather events, such as the combination of high easterly winds and low moisture levels associated with severe fires in the Pacific Northwest (Agee 1993). Some forests are more appropriately characterized at larger spatial scales by variability in intensity than by mean frequency (Morrison and Swanson 1990). If a critical threshold of intensity exists, below which late-successional species can survive once established (for example, Harmon 1984), changes in intensity may be more important ecologically than changes in frequency. In some systems in which low-intensity fires are the norm, the ecological effect of changes in frequency of low-intensity fires may need to be catalyzed by a high-intensity fire. For example, a high-intensity fire in a woodland may kill even fire-resistant dominant trees. If the fire-free interval is subsequently reduced, saplings may not survive to a fire-resistant stage (for example, Bock and Bock 1984).

Human activities—This study uses the natural vegetation types of Küchler (1964) as a basis for vegetation distribution, but in reality, much of this vegetation has been altered by humans (Klopatek and others 1979). Agricultural and urban landscapes can be presumed "lost" from the modeling database for biome-scale vegetation change, but there are many systems in which natural vegetation patterns and fire regimes coexist with human-caused disturbance.

Known effects of fire exclusion have been used above to predict reverse transitions due to increased fire frequency. These predictions can be problematic even in natural systems with no additional significant disturbance (for example, Jameson 1987). Predictions are further confounded by the presence of invading exotic species in the understory and by changes in understory composition and fuel loads due to livestock grazing (D'Antonio and Vitousek 1992, Fleischner 1994, Mack 1981). Particularly on rangelands, grazing may alter successional pathways, thereby creating a new set of successional stages distinct from those previously maintained by fire (Westoby and others 1989), and may actively disperse exotic species, which respond in various ways to increased fire frequency (Agee 1994, D'Antonio and Vitousek 1992). Grazing also reduces the fuel load of flammable grasses, and therefore the potential for frequent, low-intensity fires that preclude the establishment of shrubs in grasslands or open woodlands (Wright and Bailey 1982). It is uncertain whether predicted increases in fire frequency will overcome the inertia established and maintained by the ecological effects of grazing.

Limitations of Modeling at Large Spatial Scales

Small-scale details are lost when modeling at larger scales. O'Neill and others (1989) note two problems that make scaling upward particularly difficult: (1) higher level properties are not necessarily sums of lower level averages (the "aggregation" problem), and spatial correlations in responses by small-scale units confound unbiased estimation of averages; and (2) even if finer scales can be ignored when a system behaves in a stable manner, instabilities in low-level dynamics (for example crown fires) break the constraints imposed by larger scale properties, thereby leading to unpredictable behavior. We have tried to minimize these difficulties by aggregating systems that are relatively homogeneous with respect to fire ecology, and by modeling changes in a variable (fire frequency) that has relatively stable dynamics at the smaller scale. The "averaging" required to produce the aggregated Küchler types and the transition rules has some limitations, however.

In a steady-state model, simultaneous changes occur over areas larger than the extent of the largest fires (figs. 2 and 3). The amount of time required to accomplish this change incrementally in the real world is more commensurate with the rate of shifts in past climatic patterns than the rate of human-induced climatic changes predicted for the future. By that time, additional transitions in portions of each region or biome would be expected in response to new climatic conditions.

Distinct subsystems within the original Küchler types have been lost in the aggregation process. These subsystems are not necessarily more homogeneous than the original Küchler types combined in each of our aggregated types. For example, the lodgepole pine forests of the central Oregon Cascade Range do not appear in the Pacific Northwest in the pretransition classification (fig. 2). It is assumed that the relative homogeneity of aggregated types will persist through predicted ecological changes, but it is also reasonable to assume that biome boundaries will shift, and new patterns of homogeneity will form. Once again, the temporal scale of this process is probably shorter than that required for complete biome transitions in a steady-state model.

In transitions driven by smaller scale events, additional state variables like landscape pattern affect the process at intermediate scales. For example, fire-interval distributions are constrained by landscape heterogeneity and the spatial scale of aggregations of age classes (Baker 1989). Thus, the influence of spatial pattern confounds predictions of the effects of increased fire frequency. As noted above, adjacent smaller scale events will be correlated. This combination of intermediate-level constraints with theoretical problems in estimation magnifies the uncertainty of predictions for large spatial and temporal scales.

These limitations need to be addressed by a dynamic model that simulates the critical mechanisms or constraints at each distinct scale in the aggregation process. The transition rules developed in this paper provide an ecological perspective on possible new biome configurations and a set of constraints for large-scale, steady-state vegetation models, such as the current version of MAPSS. They also can augment the conceptual framework for monitoring changes in vegetation composition and serve as tools for the calibration of dynamic models of biome-scale vegetation change.

Acknowledgments

We thank J.K. Agee, J. Lenihan, R.P. Neilson, and D. Sandberg for helpful comments on an earlier version of this manuscript, and B. Rochefort, D. Sharkov, and S.M. Tang for help in producing the graphics.

Literature Cited

Abrams, M.D. 1986. Historical development of gallery forests in northeast Kansas. Vegetatio. 65: 29-37.

Abrams, M.D. 1992. Fire and the development of oak forests. BioScience. 42: 346-353.

Agee, J.K. 1993. Fire ecology of Pacific Northwest forests. Washington, DC: Island Press. 493 p.

Agee, J.K. 1994. Fire and weather disturbances in terrestrial ecosystems of the eastern Cascades. Gen. Tech. Rep. PNW-320. Portland, OR: U.S. Department of Agriculture, Forest Service, Pacific Northwest Research Station. 52 p.

Agee, J.K.; Dunwiddie, P. 1984. Recent forest development on Yellow Island, Washington. Canadian Journal of Botany. 62: 2074-2080.

Agee, J.K.; Huff, M.H. 1987. Fuel succession in a western hemlock/Douglas-fir forest. Canadian Journal of Forest Research. 17: 697-704.

Agee, J.K.; Smith, L. 1984. Subalpine tree establishment after fire in the Olympic Mountains, Washington. Ecology. 65: 810-819.

Anderson, H.G.; Bailey, A.W. 1980. Effects of annual burning on vegetation in the aspen parkland of east central Alberta. Canadian Journal of Botany. 58: 985-996.

Anderson, R.C.; Brown, L.E. 1986. Stability and instability in plant communities following fire. American Journal of Botany. 73: 364-368.

Andrews, P.L. 1986. BEHAVE: fire behavior prediction and fuel modeling subsystem: BURN subsystem, part 1. Gen. Tech. Rep. INT-194. Ogden, UT: U.S. Department of Agriculture, Forest Service, Intermountain Research Station. 130 p.

Arno, S.F.; Hammerly, R.P. 1984. Timberline: mountain and Arctic forest frontiers. Seattle: The Mountaineers. 304 p.

Baker, W.L. 1989. Effect of scale and spatial heterogeneity on fire-interval distributions. Canadian Journal of Forest Research. 19: 700-706.

Barrett, S.W. 1988. Fire suppression's effects on forest succession within a central Idaho wilderness. Western Journal of Applied Forestry. 3: 76-80.

Bergeron, Y.; Archambault, S. 1993. Decreasing frequency of forest fires in the southern boreal zone of Québec and its relation to global warming since the end of the 'Little Ice Age'. The Holocene. 3: 255-259.

Billings, W.D. 1969. Vegetational pattern near alpine timberline as affected by fire-snowdrift interactions. Vegetatio. 19: 191-207.

Bock, J.H.; Bock, C.E. 1984. Effects of fire on woody vegetation in the pine-grassland ecotone of the southern Black Hills. American Midland Naturalist. 112: 35-42.

Boerner, R.E.J.; Lord, T.R.; Peterson, J.C. 1988. Prescribed burning in the oak-pine forest of the New Jersey pine barrens: effects on growth and nutrient dynamics of two *Quercus* species. American Midland Naturalist. 120: 108-119.

Bonan, G.B.; Shugart, H.H. 1989. Environmental factors and ecological processes in boreal forests. Annual Review of Ecology and Systematics. 20: 1-28.

Botkin, D.B.; Janak, J.F.; Wallis, J.R. 1972. Some ecological consequences of a computer model of forest growth. Journal of Ecology. 60: 849-873.

Bradley, A.F.; Fischer, W.C.; Noste, N.V. 1992a. Fire ecology of the forest habitat types of eastern Idaho and western Wyoming. Gen. Tech. Rep. INT-290. Ogden, UT: U.S. Department of Agriculture, Forest Service, Intermountain Research Station. 92 p.

Bradley, A.F.; Noste, N.V.; Fischer, W.C. 1992b. Fire ecology of forests and woodlands in Utah. Gen. Tech. Rep. INT-287. Ogden, UT: U.S. Department of Agriculture, Forest Service, Intermountain Research Station. 128 p.

Brown, A.A.; Davis, K.P. 1973. Forest fire: control and use. New York: McGraw-Hill. 686 p.

Brubaker, L.B. 1988. Vegetation history and anticipating future vegetation change. In: Agee, J.K.; Johnson, D.R., eds. Ecosystem management for parks and wilderness. Seattle, WA: University of Washington Press: 41-61.

Bryant, W.S.; McComb, W.C.; Fralish, J.S. 1993. Oak-hickory forests. In: Martin, W.H.; Boyce, S.G.; Echternacht, A.C., eds. Biodiversity of the Southeastern United States: upland terrestrial communities. New York: Wiley and Sons: 143-202.

Buchholz, K. 1983. Initial responses of pine and oaks to wildfire in New Jersey pine barren plains. Bulletin of the Torrey Botanical Club. 110: 91-96.

Bunting, S.C. 1987. Use of prescribed burning in juniper and pinyon-juniper woodlands. In: Proceedings—pinyon-juniper conference. Gen. Tech. Rep. INT-215. Ogden, UT: U.S. Department of Agriculture, Forest Service, Intermountain Research Station: 141-151.

Cain, M.D.; Shelton, M.G. 1994. Indigenous vegetation in a southern Arkansas pine-hardwood forest after a half century without catastrophic disturbances. Natural Areas Journal. 14: 165-174.

Callaway, R.M.; Davis, F.W. 1993. Vegetation dynamics, fire and the physical environment in coastal central California. Ecology. 74: 1567-1578.

Carleton, T.J.; Maycock, P.F. 1978. Dynamics of boreal forest south of James Bay. Canadian Journal of Botany. 56: 1157-1173.

Cattelino, P.J.; Noble, I.R.; Slatyer, R.O.; Kessell, S.R. 1979. Predicting the multiple pathways of plant succession. Environmental Management. 3: 41-50.

Chappell, C.B. 1991. Fire ecology and seedling establishment in Shasta red fir (*Abies magnifica* var. *shastensis*) forests of Crater Lake National Park, Oregon. Seattle, WA: University of Washington. 133 p. M.S. thesis.

Christensen, N.L. 1985. Shrubland fire regimes and their evolutionary consequences. In: Pickett, S.T.A.; White, P.S., eds. The ecology of natural disturbance and patch dynamics. San Diego, CA: Academic Press: 85-100.

Christensen, N.L. 1988. Vegetation of the southeastern coastal plain. In: Barbour, M.G.; Billings, W.D. 1988. North American terrestrial vegetation. New York: Cambridge University Press: 317-363.

Collins, S.L. 1992. Fire frequency and community heterogeneity in tallgrass prairie vegetation. Ecology. 73: 2001-2006.

Collins, S.L.; Gibson, D.J. 1990. Effects of fire on community structure in tallgrass and mixed-grass prairie. In: Collins, S.L.; Wallace, L.L., eds. Fire in North American tallgrass prairies. Norman, OK: University of Oklahoma Press: 81-98.

Cwynar, L.C.; Speer, R.W. 1991. Reversion of forest to tundra in the central Yukon. Ecology. 72: 202-212.

D'Antonio, C.M.; Vitousek, P.M. 1992. Biological invasions by exotic grasses, the grass/fire cycle, and global change. Annual Review of Ecology and Systematics. 23: 63-87.

Desponts, M.; Payette, S. 1993. The Holocene dynamics of jack pine at its northern range limit in Quebec. Journal of Ecology. 81: 719-727.

Ehleringer, J.R.; Field, C.B. 1993. Scaling physiological processes: leaf to globe. San Diego, CA: Academic Press. 388 p.

Everett, R.L. 1987. Plant response to fire in the pinyon-juniper zone. In: Proceedings—pinyon juniper conference. Gen. Tech. Rep. INT-215. Ogden, UT: U.S. Department of Agriculture, Forest Service, Intermountain Research Station: 152-157.

Ewing, A.L.; Engle, D.M. 1988. Effects of late summer fire on tallgrass prairie microclimate and community composition. American Midland Naturalist. 1: 212-223.

Eyre, F.H. 1980. Forest cover types of the United States and Canada. Washington, DC: Society of American Foresters. 148 p.

Fahnestock, G.R.; Agee, J.K. 1983. Biomass consumption and smoke production by prehistoric and modern forest fires in western Washington. Journal of Forestry. 81: 653-657.

Finney, M.A.; Martin, R.E. 1992. Short fire intervals recorded by redwoods at Annadel State Park, CA. Madroño. 39: 251-262.

Fischer, W.C.; Bradley, A.F. 1987. Fire ecology of western Montana forest habitat types. Gen. Tech. Rep. INT-223. Ogden, UT: U.S. Department of Agriculture, Forest Service, Intermountain Research Station. 95 p.

Fischer, W.C.; Clayton, B.D. 1983. Fire ecology of Montana forest habitat types east of the Continental Divide. Gen. Tech. Rep. INT-141. Ogden, UT: U.S. Department of Agriculture, Forest Service, Intermountain Research Station. 83 p.

Fleischner, T.L. 1994. Ecological costs of livestock grazing in western North America. Conservation Biology. 8: 629-644.

Franklin, J.F. 1988. Pacific Northwest forests. In: Barbour, M.G.; Billings, W.D., eds. North American terrestrial vegetation. New York: Cambridge University Press: 103-130.

Furyaev, V.V.; Wein, R.W.; MacLean, D.A. 1983. Fire influences in *Abies* dominated forests. In: Wein, R.W.; MacLean, D.A., eds. The role of fire in Northern circumpolar ecosystems. New York: Wiley and Sons: 221-234.

Gibson, D.J.; Hulbert, L.C. 1987. Effects of fire, topography and year-to-year climatic variation on species composition in tallgrass prairie. Vegetatio. 72: 175-185.

Green, D.G. 1989. Simulated effects of fire, dispersal and spatial pattern on competition within forest mosaics. Vegetatio. 82: 139-153.

Grimm, E.C. 1984. Fire and other factors controlling the big woods vegetation of Minnesota in the mid nineteenth century. Ecological Monographs. 54: 291-311.

Harmon, M.E. 1984. Survival of trees after low-intensity surface fires in Great Smoky Mountains National Park. Ecology. 65: 796-802.

Hartnett, D.C.; Krofta, D.M. 1989. Fifty-five years of post-fire succession in a southern mixed hardwood forest. Bulletin of the Torrey Botanical Club. 116: 107-113.

Heinselman, M.L. 1973. Fire in the virgin forests of the Boundary Waters Canoe Area, Minnesota. Quaternary Research. 3: 329-382.

Heinselman, M.L. 1981. Fire intensity and frequency as factors in the distribution and structure of northern ecosystems. In: Mooney, H.A.; Bonnicksen, T.M. [and others], eds. Fire regimes and ecosystem properties. Gen. Tech. Rep. WO-26. Washington, DC: U.S. Department of Agriculture, Forest Service: 7-57.

Host, G.E.; Pregitzer, K.S. [and others]. 1987. Landform mediated differences in successional pathways among upland forest ecosystems in northwestern Lower Michigan. Forest Science. 33: 445-457.

Howe, H.F. 1994. Managing species diversity in tallgrass prairie: assumptions and implications. Conservation Biology. 8: 691-704.

Huff, M.H. 1984. Post-fire succession in the Olympic Mountains, Washington: forest vegetation, fuels and avifauna. Seattle, WA: University of Washington. 240 p. Ph.D. dissertation.

Humphrey, R.R. 1974. Fire in the deserts and desert grassland of North America. In: Kozlowski, T.T.; Ahlgren, C.E., eds. Fire and ecosystems. New York: Academic Press: 365-400.

Jameson, D.A. 1987. Climax or alternative steady states in woodland ecology. In: Proceedings—pinyon juniper conference. Gen. Tech. Rep. INT-215. Ogden, UT: U.S. Department of Agriculture, Forest Service, Intermountain Research Station: 9-13.

Kauffman, J.B.; Martin, R.E. 1987. Effects of fire and fire suppression on mortality and mode of reproduction of California black oak. In: Plumb, T.R.; Pillsbury, N.H., eds. Proceedings of the symposium on multiple use management of California's hardwood resources; 1986 November 12-14, San Luis Obispo, CA. Gen. Tech. Rep. PSW-100. Berkeley, CA: U.S. Department of Agriculture, Forest Service, Pacific Southwest Research Station: 122-126.

Keane, R.E.; Arno, S.F.; Brown, J.K. 1989. FIRESUM: an ecological process model for fire succession in western conifer forests. Gen. Tech. Rep. INT-266. Ogden, UT: U.S. Department of Agriculture, Forest Service, Intermountain Research Station. 76 p.

Keane, R.E.; Reinhardt, E.D.; Brown, J.K. 1994. FOFEM: a first order fire effects model for predicting the immediate consequences of wildland fire in the United States. In: Proceedings of the 12th conference on fire and forest meteorology; 1993 October 26-28; Jekyll Island, GA. Publ. 94-02. Bethesda, MD: Society of American Forsters: 628-631.

Keeley, J.E.; Keeley, S.C. 1988. Chaparral. In: Barbour, M.G.; Billings, W.D., eds. North American terrestrial vegetation. New York: Cambridge University Press: 165-208.

Keown, L.D. 1977. Interim report: Blacktail Hills prescribed fire project, implementation and results. Great Falls, MT: U.S. Department of Agriculture, Forest Service, Lewis and Clark National Forest. 9 p.

Kercher, J.R.; Axelrod, M.C. 1984. A process model of fire ecology and succession in a mixed conifer forest. Ecology. 65: 1725-1742.

Kessell, S.R. 1976. Gradient modeling: a new approach to fire modeling and wilderness resource management. Journal of Environmental Management. 1: 39-48.

King, A.W.; Emanuel, W.R.; O'Neill, R.V. 1990. Linking mechanistic models of tree physiology with models of forest dynamics: problems of temporal scale. In: Dixon, R.K.; Meldahl, R.S.; Ruark, G.A.; Warren, W.G., eds. Process modeling of forest growth responses to environmental stress. Portland, OR: Timber Press: 241-248.

Klopatek, J.M.; Olson, R.J.; Emerson, C.J. 1979. Land use conflicts with natural vegetation in the United States. Environmental Conservation. 6: 191-199.

Küchler, A.W. 1964. Potential natural vegetation of the coterminous United States. New York: American Geographical Society. 116 p.

Landhäusser, S.M.; Wein, R.W. 1993. Postfire vegetation recovery and tree establishment at the Arctic treeline: climate change—vegetation response hypotheses. Journal of Ecology. 81: 665-672.

Little, R.L.; Peterson, D.L.; Conquest, L.L. 1994. Regeneration of subalpine fir (*Abies lasiocarpa*) following fire: effects of climate and other factors. Canadian Journal of Forest Research. 24: 934-944.

Lorimer, C.G. 1985. The role of fire in the perpetuation of oak forests. In: Johnson, J.E., ed. Challenges in oak management and utilization. Madison WI: University of Wisconsin: 8-25.

Mack, R.N. 1981. Invasion of *Bromus tectorum* L. into western North America: an ecological chronicle. Agro-Ecosystems. 7: 145-165.

Marsden, M.A. 1983. Modeling the effect of wildfire frequency on forest structure and succession in the northern Rocky Mountains. Journal of Environmental Management. 16: 45-62.

Martin, S.C. 1975. Ecology and management of southwestern semidesert grass-shrub ranges: the status of our knowledge. Res. Pap. RM-156. Moscow, ID: U.S. Department of Agriculture, Forest Service, Rocky Mountain Research Station. 39 p.

Miller, R.F.; Wigand, P.E. 1994. Holocene changes in semi-arid pinyon-juniper woodlands. BioScience. 44: 465-474.

Morneau, C.; Payette, S. 1989. Postfire lichen-spruce woodland recovery at the limit of the boreal forest in northern Quebec. Canadian Journal of Botany. 67: 2770-2782.

Morrison, P.; Swanson, F.J. 1990. Fire history and pattern in a Cascade Range landscape. Gen. Tech. Rep. PNW-254. Portland, OR: U.S. Department of Agriculture, Forest Service, Pacific Northwest Research Station. 77 p.

Muir, P.S. 1993. Distubance effects on structure and tree species composition of *Pinus contorta* forests in western Montana. Canadian Journal of Forest Research. 23: 1617-1625.

Myers, R.L. 1985. Fire and the dynamic relationship between Florida sandhill and sand pine scrub vegetation. Bulletin of the Torrey Botanical Club. 112: 241-252.

Neilson, R.P. 1992. Toward a rule-based biome model. Landscape Ecology. 7: 27-43.

Nowacki, G.J.; Abrams, M.D.; Lorimer, C.G. 1990. Composition, structure and historical development of northern red oak stands along an edaphic gradient in north central Wisconsin. Forest Science. 36: 276-292.

O'Neill, R.V.; Johnson, A.R.; King, A.W. 1989. A hierarchical framework for the analysis of scale. Landscape Ecology. 3: 193-205.

Orwig, D.A.; Abrams, M.D. 1994. Land use history (1720-1992), composition and dynamics of oak-pine forests within the Piedmont and coastal plain of northern Virginia. Canadian Journal of Forest Research. 24: 1216-1225.

Pastor, J.; Mladenoff, D.J. 1992. The Southern boreal/Northern hardwood forest border. In: Shugart, H.H.; Leemans, R.; Bonan, G., eds. A systems analysis of the global boreal forest. New York: Cambridge University Press: 216-240.

Payette, S. 1992. Fire as a controlling process in the North American boreal forest. In: Shugart, H.H.; Leemans, R.; Bonan, G., eds. A systems analysis of the global boreal forest. New York: Cambridge University Press: 144-169.

Payette, S. 1993. The range limit of boreal tree species in Quebec-Labrador: an ecological and palaeoecological interpretation. Review of Palaeobotany and Palynology. 79: 7-30.

Payette, S.; Morneau, C.; Sirois, L.; Desponts, M. 1989. Recent fire history of the northern Quebec biomes. Ecology. 70: 656-673.

Peet, R.K. 1981. Forest vegetation of the Colorado Front Range. Vegetatio. 45: 3-75.

Peterson, D.L.; Ryan, K.C. 1986. Modeling post-fire conifer mortality for long range planning. Environmental Management. 10: 797-808.

Price, C.; Rind, D. 1994. The impact of a 2 x CO_2 climate on lightning-caused fires. Journal of Climate. 7: 1484-1494.

Qu, J.; Omi, P.N. 1994. Potential impacts of global climate changes on wildfire activity in the USA. In: Proceedings of the 12th conference on fire and forest meteorology; 1993 October 26-28; Jekyll Island, GA. Publ. 94-02. Bethesda, MD: Society of American Foresters: 85-92.

Quaterman, E.; Keever, C. 1962. Southern mixed hardwood forest: climax in the southeastern coastal plain. Ecological Monographs. 32: 167-185.

Reed, L.J.; Sugihara, N.G. 1987. Northern oak woodlands: ecosystem in jeopardy or is it already too late? In: Plumb, T.R.; Pillsbury, N.H., eds. Proceedings of the symposium on multiple use management of California's hardwood resources; 1986 November 12-14; San Luis Obispo, CA. Gen. Tech. Rep. PSW-100. Berkeley, CA: U.S. Department of Agriculture, Forest Service, Pacific Southwest Research Station: 59-63.

Risser, P.G. 1990. Landscape processes and the vegetation of the North American grassland. In: Collins, S.L.; Wallace, L.L., eds. Fire in North American tallgrass prairies. Norman, OK: University of Oklahoma Press: 133-146.

Rochefort, R.M.; Little, R.L.; Woodward, A.; Peterson, D.L. 1994. Changes in subalpine tree distribution in western North America: a review of climate and other causal factors. The Holocene. 4: 89-100.

Rostlund, E. 1957. The myth of a natural prairie belt in Alabama: an interpretation of historical records. Annals of American Geography. 47: 392-411.

Rothermel, R.C. 1972. A mathematical model for predicting fire spread in wildland fuels. Res. Pap. INT-115. Ogden, UT: U.S. Department of Agriculture, Forest Service, Intermountain Research Station. 40 p.

Rothermel, R.C. 1991. Predicting behavior and size of crown fires in the northern Rocky Mountains. Res. Pap. INT-438. Ogden, UT: U.S. Department of Agriculture, Forest Service, Intermountain Research Station. 46 p.

Running, S.W.; Hunt, R. 1993. Generalization of a forest ecosystem process model for other biomes, biome-BGC, and an application for global-scale models. In: Ehleringer, J.R.; Field, C.B., eds. Scaling physiological processes: leaf to globe. San Diego, CA: Academic Press: 141-158.

Ryan, K.C.; Reinhardt, E.D. 1988. Predicting postfire mortality of seven western conifers. Canadian Journal of Forest Research. 18: 1291-1297.

Schneider, S.H. 1989. Global warming. San Francisco, CA: Sierra Club Books. 317 p.

Shugart, H.H.; Prentice, I.C. 1992. Individual tree based models of forest dynamics and their application in global change research. In: Shugart, H.H.; Leemans, R.; Bonan, G., eds. A systems analysis of the global boreal forest. New York: Cambridge University Press: 313-333.

Sirois, L.; Payette, S. 1991. Reduced postfire tree regeneration along a boreal forest-forest-tundra transect in northern Quebec. Ecology. 72: 619-627.

Slatyer, R.O.; Noble, I.R. 1992. Dynamics of montane treelines. In: Hansen, A.J.; di Castri, F., eds. Landscape boundaries: consequences for biotic diversity and ecological flows: Ecological Studies 92. New York: Springer-Verlag: 346-359.

St. Pierre, H.; Gagnon, R.; Bellefleur, P. 1991. Post-fire spatial distribution of black spruce and jack pine regeneration in the boreal forest of the Ashuapmushan Fauna Reserve, Quebec. Canadian Journal of Botany. 69: 717-721.

Stauffer, D. 1985. Introduction to percolation theory. London: Taylor and Francis. 124 p.

Strauss, D.; Bednar, L.; Mees, R. 1989. Do one percent of forest fires cause ninety-nine percent of the damage? Forest Science. 35: 319-328.

Taylor, A.H. 1993. Fire history and structure of red fir (Abies magnifica) forests, Swain Mountain Experimental Forest, Cascade Range, northeastern California. Canadian Journal of Forest Research. 23: 1672-1678.

Tester, J.R. 1989. Effects of fire frequency on oak savanna in east-central Minnesota. Bulletin of the Torrey Botanical Club. 116: 134-144.

Turner, M.G.; Romme, W.H. 1994. Landscape dynamics in crown fire ecosystems. Landscape Ecology. 9: 59-77.

Vale, T.R. 1981. Tree invasion of mature montane meadows in Oregon. American Midland Naturalist. 105: 61-69.

van Wagtendonk, J.W. 1985. Fire suppression effects on fuels and succession in short-fire-interval wilderness ecosystems. In: Lotan, J.E.; Kilgore, B.M.; Fischer, W.C.; Mutch, R.W., eds. Proceedings, symposium and workshop on wilderness fire; 1983 November 15-18; Missoula, MT. Gen. Tech. Rep. INT-182. Ogden, UT: U.S. Department of Agriculture, Forest Service, Intermountain Research Station: 119-126.

Vose, J.M.; Clinton, B.D.; Swank, W.T. 1994. Fire, drought, and forest manage-ment influences on pine/hardwood ecosystems in the southern Appalachians. In: Proceedings of the 12th conference on fire and forest meteorology; 1993 October 26-28; Jekyll Island, GA. Publ. 94-02. Bethesda, MD: Society of American Foresters: 85-92.

Westoby, M.; Walker, B.; Noy-Meir, I. 1989. Opportunistic management for rangelands not at equilibrium. Journal of Range Management. 42: 266-274.

Woodward, F.I. 1987. Climate and plant distribution. London: Cambridge University Press. 174 p.

Woodward, F.I.; McKee, I.F. 1991. Vegetation and climate. Environmental International. 17: 535-546.

Woodward, P.M. 1994. Fire history and tree establishment in a subalpine forest in eastern Washington. In: Proceedings of the 12th conference on fire and forest meteorology; 1993 October 26-28; Jekyll Island, GA. Publ. 94-02. Bethesda, MD: Society of American Foresters: 209-222.

Wright, H.A.; Bunting, S.C.; Neuenschwander, L.F. 1976. Effect of fire on honey mesquite. Journal of Range Management. 29: 467-471.

Wright, H.A.; Bailey, A.W. 1982. Fire ecology: United States and Southern Canada. New York: Wiley and Sons. 501 p.

Zedler, P.H.; Gautier, C.R.; McMaster, G.S. 1983. Vegetation change in response to extreme events: the effects of a short interval between fires in California chaparral and coastal shrub. Ecology. 64: 809-818.

Appendix

Table 6—Descriptions and key species for aggregated Küchler types

Küchler type	Description	Key species
Spruce-hemlock	Southeast Alaska to north Oregon Coast Range, fire currently not a major agent of disturbance.	*Picea sitchensis, Tsuga heterophylla, Thuja plicata*
Hemlock/Douglas-fir	Western slope of the Cascade Range or western slope, Rockies, infrequent but severe fires. Low to mid elevations.	*Pseudotsuga menziesii, Tsuga heterophylla, Thuja plicata*
Silver fir/Douglas-fir	Mid to upper elevations in the Cascade Range and Olympics between subalpine and hemlock/Douglas-fir.	*Abies amabilis, Pseudotsuga menziesii*
W. fir-spruce	High-elevation forests. Decreasing moisture-successional gradient from west to east in the Pacific Northwest, corresponding to increased fire frequency.	*Abies lasiocarpa, Tsuga mertensiana, Pinus albicaulis, P. contorta, Picea engelmannii, A. magnifica*
Mixed conifer	Includes the Sierra Nevada, and northern California Coast Range into Oregon, and high elevation in southern California.	*Abies concolor, Pinus lambertiana, P. ponderosa, Pseudotsuga menziesii, Arbutus menziesii, Quercus chrysolepis*
Redwood	California coast and southern Sierra Nevada, forests dominated by coast redwood.	*Sequoia sempervirens, Pseudotsuga menziesii*
Lodgepole pine	Inland, Rocky Mountains, Oregon Cascade Range and high elevations in Western United States. Severe stand-replacing fires.	*Pinus contorta, Abies lasiocarpa, Picea engelmannii*
Pine-cypress	Narrow range along California coast from Monterey to San Diego. Many isolated stands.	*Cupressus goveniana, Pinus contorta* var. *contorta*
Ponderosa pine	SW Oregon, California mountains, eastern slope of the Cascade Range. Also Rocky Mountains, Black Hills, Arizona mountains. Low-severity, high-frequency fire.	*Pinus ponderosa*
Douglas-fir	Cascade Range and inland mountains. Drier than hemlock/Douglas-fir, can include grand fir or other conifers, either as seral or potential climax species.	*Pseudotsuga menziesii, Abies grandis, Picea pungens* (inland states)
Cedar-hemlock-pine	Northern Rocky Mountains western slope.	*Thuja plicata, Tsuga heterophylla, Pinus monticola, Pseudotsuga menziesii, Pinus ponderosa* (low elevations)
Great Basin pine	Bristlecone and limber pine systems, spanning lower treeline elevations, in Great Basin mountain ranges. High-elevation ecotone with alpine tundra, lower with sage-steppe.	*Pinus aristata, P. flexilis*
Pinyon-juniper	Dry Western United States, below oak woodlands, fire adapted somewhat, can coexist with frequent fire in more mesic systems. Fire will exclude in drier systems.	*Pinus edulis, Juniperus occidentalis, Artemisia tridentata*
Alder-ash	MAPSS has this as wetland, but could be pioneering vegetation on spruce-hemlock sites after disturbance. In Pacific Northwest, much in Willamette Valley.	*Alnus rubra, Fraxinus latifolia, Populus trichocarpa*

W. oakwoods	Oak woodlands in Oregon and California. Fire exclusion leads to invasion and dominance by conifers.	*Quercus garryana, Q. agrifolia, Q. chrysolepis*
Mesquite savanna	Warm and dry-mesic, where mesquite is important shrub. Southern Arizona, Texas, and Oklahoma.	*Prosopis juliflora*
Oak-juniper	Warm regions in Southern United States, somewhat adapted to fire.	*Juniperus ashei, Quercus virginiana, J. virginiana, Q. stellata, Andropogon scoparius* (E). *J. deppeana, Q. emoryi* (W)
Chaparral	Mainly in California. Includes some oaks, also montane chaparral in Utah and Colorado.	*Arctostaphylos* spp., *Ceanothus* spp., *Cercocarpus ledifolius, Quercus gambelii*
Great Basin shrubland	Sagebrush, blackbrush, "shrub-steppe" in Inter-mountain West.	*Artemisia tridentata, Coleogyne ramossisima, Sarcobatus vermiculatus*
Desert shrubland	Sparse shrubs, and not much grass. Similar to desert grasslands and chaparral in California and Southwest in being fire adapted.	*Larrea divaricata, Andropogon scoparius*
Desert grassland	Very dry grasslands, not enough continuous shrubs to burn. Southwest including west Texas.	*Larrea divaricata, Cercidium microphyllum, Bouteloua eriopoda*
Shortgrass prairie mixed-grass prairie tallgrass prairie	Tall and shortgrass prairies in central and Western United States, adapted to frequent fires, without which trees will invade at ecotones. Short → tall grasses predominate as moisture increases west to east.	*Andropogon gerardii* (tallgrass), *Agropyron spicatum* and *Festuca idahoensis* (mixed), *Bouteloua gracilis* (shortgrass)
Grassland-wetland	Nonforested wetlands in Southwest and Southeast, including tule marshes, cordgrass, and everglades.	*Carex* spp., *Gerardia maritima, Mariscus jamaicensis*
Oak savanna	Prairie borders in the eastern Great Plains, also eastern and Gulf States. Fire-adapted, fire exclusion leads to invasion of shade-tolerant deciduous forest species.	*Quercus macrocarpa, Q. virginiana, Andropogon gerardii, A. scoparius*
Blackbelt (oak-gum-cypress)	Oak-gum-cypress forest-savanna, with some tallgrass prairie, in Alabama and Mississippi.	*Juniperus virginiana, Liquidambar styraciflua, Quercus stellata*
SE wetland (forested)	Cypress savanna (Florida), mangrove, southern flood plains and pocosins.	*Aristida* spp., *Avicennia nitida, Nyssa aquatica, Pinus serotina*
E. spruce-fir	High-elevation forests in Southeastern United States and conifer forests of Eastern Canada and Maine. Fires are infrequent.	*Abies fraseri* and *Picea rubens* (SE), *A. balsamea* and *P. rubens* (NE)
Conifer bog	Tamarack-black spruce dominated bogs in south-central Canada and Lake States.	*Picea mariana, Larix laricina, Thuja occidentalis*
White-red-jack pine	Küchler category Great Lakes pine. Pine is early successional but dominates on these sites due to frequent fires.	*Pinus banksiana, P. resinosa, P. strobus*

Table 6—Descriptions and key species for aggregated Küchler types (continued)

Küchler type	Description	Key species
N. flood plain	Elm-ash-cottonwood bottomland forests in central United States	*Populus deltoides, Salix nigra, Ulmus americana*
E. deciduous	Eastern United States, late-successional deciduous species, resistant to fire once trees are mature and the canopy is closed (for example, Minnesota Big Woods). Includes "mixed mesophytic" forest in central Eastern United States	*Acer saccharum, Fagus grandifolia, Tilia americana, Quercus rubra, Carya* spp.
Conifer-hardwoods	Transition between boreal forest and E. deciduous, complex successional dynamics, mediated by fire and herbivory. Southern central Canada, upper Great Lakes States, in north woods in Maine these forests are spruce-fir mixed with hardwoods.	*Acer saccharum, Fagus grandifolia, Picea rubens, Tsuga canadensis, Abies balsamea, Betula allegheniensis*
Oak-hickory	Central Eastern United States, south of E. deciduous forest and more adapted to frequent fire.	*Quercus* spp., *Carya* spp.
Oak-pine	Eastern United States down to subtropical zone in Flordia. Pines are fire adapted, and percentage of pine is related to importance of fire.	*Pinus rigida* (NE), *P. taeda, P. echinata* (SE)
S. mixed forest	Beech, gum, pine, and oak in Southeast. Succession proceeds from pine to hardwoods, reversed by fire.	*Fagus grandifolia, Liquidambar styraciflua, Pinus elliottii, Quercus alba*
SE pine	Originally longleaf pine, many of these forests have been cut and planted with loblolly and slash pine. The system is fire driven.	*Pinus* spp.
Arctic-alpine tundra	Northern Canada and Alaska north of treeline, and elevations above treeline in Cascade Range, Rocky Mountains, and Sierra Nevada. Meadows are invaded by subalpine tree species in favorable conditions.	*Lupinus* spp., *Carex* spp., *Festuca* spp.
Desert	United States Great Basin and Southwest. Not much fire, little to burn. Moisture-temperature driven though Great Basin high shrubland will burn. In extreme desert (for example, Mojave) fires cannot spread.	*Opuntia* spp.
Boreal woodland	South of Arctic tundra, north of continuous forest line, lichen-black spruce woodlands, peat bogs.	*Picea mariana*
Boreal forest	Stretching across Canada, increasing moisture gradient from west to east. Stand-replacing fires are common.	*Pinus banksiana, Picea glauca, P. mariana, Abies balsamea, Populus tremuloides, Betula papyrifera*
Aspen-birch	Within the conifer-hardwoods zone, site dominated by aspen and birch.	*Populus tremuloides, Betula papyrifera*
Aspen parkland	Ecotone between boreal forest and northern prairie. Increasing fire and warmth along gradient north to south.	*Populus tremuloides*

Table 7—Correspondences among aggregated Küchler, MAPSS, and SAF vegetation types

Spruce-hemlock	Forest mixed warm (evergreen)	Sitka spruce (223), western hemlock (224), western hemlock-Sitka spruce (225), western redcedar (228)
Hemlock/Douglas-fir	Forest mixed warm (evergreen)	Pacific Douglas-fir (229), Douglas-fir-western hemlock (230), Port-Orford-cedar (231)
Silver fir/Douglas-fir	Forest evergreen needle	Coastal true fir-hemlock (226)
W. fir-spruce	Forest evergreen needle-tree savanna evergreen needle	Mountain hemlock (205), Engelmann spruce-subalpine fir (206), red fir (207), whitebark pine (208), California mixed subalpine (256), aspen (217)
Mixed conifer	Forest evergreen needle-tree savanna mixed warm	Sierra Nevada mixed conifer (243), Pacific ponderosa pine Douglas-fir (245), Jeffrey pine (247), knobcone pine (248), white fir (211)
Redwood	Forest mixed warm (evergreen)	Redwood (232)
Lodgepole pine	Forest evergreen needle	Lodgepole pine (218)
Pine-cypress	Forest mixed warm (evergreen)	(No corresponding type.)
Ponderosa pine	Tree savanna evergreen needle	Pacific ponderosa pine (245), interior ponderosa pine (237)
Douglas-fir	Forest evergreen needle	Interior Douglas-fir (210), Douglas-fir tanoak Pacific madrone (234), grand fir (213), blue spruce (216), western larch (212)
Cedar-hemlock-pine	Forest evergreen needle	Western redcedar-western hemlock (227), western white pine (215)
Great Basin pine	Tree savanna evergreen needle	Bristlecone pine (209), limber pine (219)
Pinyon-juniper	Tree savanna evergreen needle	Rocky mountain juniper (220), pinyon-juniper (239), western juniper (238), Arizona cypress (240)
Alder-ash	Wetland	Red alder (221), black cottonwood-willow (222)
W. oakwoods-oak savanna	Tree savanna deciduous broadleaf, tree savanna mixed warm	Oregon white oak (233), western live oak (241), California black oak (246), canyon live oak (249), blue oak-digger pine (250), California coast live oak (255) post oak-blackjack oak (40), bur oak (42), live oak (89)
Mesquite	Shrub savanna mixed warm	Mesquite
Oak-juniper	Tree savanna mixed warm	(No corresponding type.)
Chaparral	Chaparral	Nonforested types not included in SAF classification
Great Basin shrubland	Shrub savanna evergreen needle	

chler type	MAPSS type	SAF type
ackbelt (oak-gum-ypress)	Tree savanna mixed warm	Swamp chestnut oak-cherrybark oak (91), sweetgum-willow oak (92), live oak (89), willow oak-water oak-diamondleaf oak (88)
wetland (forested)	Wetland	Baldcypress (101), Baldcypress-tupelo (102), water tupelo-swamp tupelo (103), sweetbay-swamp tupelo-redbay (104), Atlantic white cedar (97), pondcypress (100), mangrove (106), tropical hardwoods (105), pond pine (98)
spruce-fir	Forest evergreen needle	Red spruce (32), red spruce-balsam fir (33), red spruce-Fraser fir (34), red spruce-yellow birch (30), paper birch-red spruce-balsam fir (35), northern white cedar (37)
nifer bog	Forest evergreen needle	Black spruce-tamarack (13), tamarack (38)
ite-red-jack pine	Forest evergreen needle	Jack pine (1), red pine (15), eastern white pine (21), white pine-northern red oak-red maple (20), white pine-chestnut oak (51)
flood plain	Wetland	Black ash-American elm-red maple (39), river birch-sycamore (61), silver maple-American elm (62), cottonwood (63), pin oak-sweetgum (65), sugarberry-American elm-green ash (93), sycamore-sweetgum-American elm (94), black willow (95), overcup oak-water hickory (96)
deciduous	Forest hardwood cool	Sugar maple (27), sugar maple-beech-yellow birch (25), sugar maple-basswood (26), black cherry-maple (28), red spruce-sugar maple-beech (31), yellow poplar (57), yellow poplar-eastern hemlock (58), beech-sugar maple (60), red maple (108), hawthorn (109)
conifer-hardwoods	Forest mixed cool	White pine-hemlock (22), eastern hemlock (23), hemlock-yellow birch (24)
k-hickory	Forest deciduous broadleaf	Post oak-blackjack oak (40), bur oak (42), bear oak (43), chestnut oak (44), white oak-black oak-northern red oak (52), white oak (53), black oak (110), northern red oak (55), black locust (50), yellow poplar-white oak-northern red oak (59), sassafras-persimmon (64), eastern redcedar (46), shortleaf pine (75)
k-pine	Forest deciduous broadleaf	Pitch pine (45), longleaf pine-scrub oak (71), shortleaf pine-oak (76), Virginia pine-oak (78), loblolly pine-hardwood (82), slash pine-hardwood (85), Virginia pine (79)
mixed forest	Forest mixed warm (deciduous)	Sweetgum-yellow poplar (87) note mixed forests are difficult to categorize by cover type classifications
pine	Forest savanna dry tropical	Sand pine (69), longleaf pine (70), longleaf pine-slash pine (83), loblolly pine (81), loblolly pine-shortleaf pine (80), slash pine (84), south Florida slash pine (111)
eal forest	Forest evergreen needle (taiga)	Balsam fir (5), black spruce (12), white spruce (107)
pen-birch	Forest hardwood cool	Aspen (16), pin cherry (17), paper birch (18), gray birch-red maple (19)

Table 7—Correspondences among aggregated Küchler, MAPSS, and SAF vegetation types (continued)

Boreal woodland	Taiga tundra	
Arctic-alpine tundra	Tundra	
Desert	Desert extreme	
Shortgrass prairie	Grass prairie short	—
Mixed-grass prairie	Grass prairie mixed	
Tallgrass prairie	Grass prairie tall	
Desert grassland	Semidesert grassland	
Desert shrubland	Shrub savanna mixed warm	—

— = No SAF classification for these nonforested type.

Table 8 — List of species mentioned in the text

Common name	Scientific name
Balsam fir	*Abies balsamea* (L.) Mill
Basswood	*Tilia americana* (L.)
Black mangrove	*Avicennia nitida* (L.)
Black spruce	*Picea mariana* (Mill) B.S.P.
Big bluestem	*Andropogon gerardii* Vitman.
Bristlecone pine	*Pinus aristata* Engelm.
Coast redwood	*Sequoia sempervirens* (D. Don) Endl.
Douglas-fir	*Pseudotsuga menziesii* (Mirb.) Franco
Eastern hemlock	*Tsuga canadensis* (L.) Carr.
Eastern white pine	*Pinus strobus* Dougl. ex D. Don
Grand fir	*Abies grandis* (Dougl. ex D. Don) Lindl.
Jack pine	*Pinus banksiana* Lamb.
Limber pine	*Pinus flexilis* James
Lodgepole pine	*Pinus contorta* Dougl. ex Loud.
Mesquite	*Prosopis juliflora* (SW.) DC.
Oregon white oak	*Quercus garryana* Dougl.
Pacific silver fir	*Abies amabilis* Dougl. ex Forbes
Paper birch	*Betula papyrifera* Marsh.
Pinyon pine	*Pinus edulis* Engelm.
Ponderosa pine	*Pinus ponderosa* Dougl. ex Laws.
Quaking aspen	*Populus tremuloides* Michx.
Red fir	*Abies magnifica* A. Murr.
Red pine	*Pinus resinosa* Ait.
Red mangrove	*Rhizophora mangle* (L.)
Sitka spruce	*Picea sitchensis* (Bong.) Carr.
Subalpine fir	*Abies lasiocarpa* (Hook.) Nutt.
Western larch	*Larix occidentalis* Nutt.
Western hemlock	*Tsuga heterophylla* (Raf.) Sarg.
Western white pine	*Pinus monticola* Dougl. ex D. Don
White fir	*Abies concolor* (Gord. & Glend.) Lindl. ex Hildebr.

McKenzie, Donald; Peterson, David L.; Alvarado, Ernesto. 1996. Predicting the
effect of fire on large-scale vegetation patterns in North America. Res. Pap.
PNW-RP-489. Portland, OR: U.S. Department of Agriculture, Forest Service,
Pacific Northwest Research Station. 38 p.

Changes in fire regimes are expected across North America in response to anticipated global climatic changes. Potential changes in large-scale vegetation patterns
are predicted as a result of altered fire frequencies. A new vegetation classification
was developed by condensing Küchler potential natural vegetation types into aggregated types that are relatively homogeneous with respect to fire regime. Transition
rules were developed to predict potential changes from one vegetation type to another
because of increased fire frequency. In general, vegetation currently associated with
warmer or drier climates could replace existing vegetation in most biomes. Exceptions
are subalpine forests and woodlands at the northern Arctic treeline, which are predicted to become treeless. The transition rules provide an ecological perspective on
possible new configurations of vegetation types, a set of constraints for steady-state
models, and a potential method of calibration for dynamic models of large-scale vegetation change.

Keywords: Biome scale, Küchler types, fire effects, transition rules, vegetation change.

The Forest Service of the U.S. Department of
Agriculture is dedicated to the principle of multiple
use management of the Nation's forest resources
for sustained yields of wood, water, forage, wildlife,
and recreation. Through forestry research,
cooperation with the States and private forest
owners, and management of the National Forests
and National Grasslands, it strives—as directed by
Congress—to provide increasingly greater service
to a growing Nation.

The United States Department of Agriculture (USDA)
prohibits discrimination in its programs on the basis
of race, color, national origin, sex, religion, age, disability,
political beliefs, and marital or familial status. (Not all
prohibited bases apply to all programs.) Persons with
disabilities who require alternative means of communication
of program information (Braille, large print, audiotape, etc.)
should contact the USDA Office of Communications at
(202) 720-2791.

To file a complaint, write the Secretary of Agriculture,
U.S. Department of Agriculture, Washington, DC 20250,
or call (202) 720-7327 (voice), or (202) 720-1127 (TDD).
USDA is an equal employment opportunity employer.

Pacific Northwest Research Station
333 S.W. First Avenue
P.O. Box 3890
Portland, Oregon 97208-3890

CPSIA information can be obtained
at www.ICGtesting.com
Printed in the USA
BVHW091737021118
531990BV00019B/1042/P

9 781528 422307